Portage Public Library

THE FUTURE OF FARMING
UNDERSTANDING GLOBAL ISSUES

Published by Smart Apple Media
1980 Lookout Drive
North Mankato, Minnesota 56003
USA

This book is based on *The Future of Farming: What Price the Food We Eat?*
Copyright ©1997 Understanding Global Issues Ltd., Cheltenham, England.

This edition copyright ©2003 WEIGL EDUCATIONAL PUBLISHERS LTD. All rights reserved. No part of this publication may be reproduced, stored in a retrieval system, or transmitted in any form or by any means, electronic, mechanical, photocopying, recording, or otherwise, without the prior written permission of the publisher.

Library of Congress Cataloging-in-Publication Data

The future of farming / edited by Jared Keen.
   p. cm. -- (Understanding global issues)
Summary: Explores the future of farming on a global scale, including the environmental, political, social, and economic implications of farming.
 ISBN 1-58340-170-9 (hardcover : alk. paper)
 1. Agriculture--Juvenile literature. 2. Agricultural innovations--Juvenile literature. [1. Agriculture. 2. Agricultural innovations.] I. Keen, Jared. II. Series.
S519 .F87 2002

2001008444

Printed in Malaysia
2 4 6 8 9 7 5 3 1

EDITOR Michael Lowry   **COPY EDITOR** Nicole Bezic King
TEXT ADAPTATION Jared Keen   **DESIGNER** Terry Paulhus
PHOTO RESEARCHER Tina Schwartzenberger

Contents

Introduction 5

Food Supply and Demand 6

Industrialized Agriculture 11

The Global Harvest 16

Livestock Farming 20

Map of Cereal Production 26

Charting the World's Food Production 28

Farm Trade and Subsidies 30

Who's Afraid of Biotech? 35

Sustainable Agriculture 40

Time Line of Events 46

Concept Web 48

Quiz 50

Internet Resources 52

Further Reading 53

Glossary 54

Index 55

Introduction

The science of agriculture is constantly changing. The latest revolution in farming has emerged from cutting-edge genetic research, so-called **biotechnology**. Biotechnology may soon replace current seeds, fertilizers, and pesticides with new and more efficient means of food production. However, efficient farming methods are not necessarily the best.

At the same time, there is growing support for systems of food production that do not rely on chemicals and factory farming. While **organic farming** is becoming increasingly popular, it still only accounts for less than five percent of farm output in most **developed countries**.

Since 1945, countries all over the world have focused on "industrializing" agriculture in order to increase the production of food. This mass production of grains, potatoes, and vegetables involves large single-crop fields liberally sprinkled with fertilizers and pesticides. Mass production of meat, eggs, and milk means putting millions of cows, sheep, pigs, and chickens through factory farms. Food production has increased as a result of industrialized agricultural methods, but so have the environmental costs. In light of these costs, the governmental policies of some countries are beginning to shift in support of **sustainable agriculture** rather than just increased production.

> *As a result of industrialized agriculture, food production has increased, but so have the environmental costs.*

Developed countries have more than enough to eat, despite the fact that they use wasteful methods of turning plants into food. People first began raising cattle so they could turn grass, which is inedible for humans, into protein-rich food. Today, most farmers have switched to feeding livestock with grain, which produces fatter animals more quickly. However, feeding livestock grain puts pressure on world food supplies. For example, it takes between 33 and 55 pounds (15–25 kg) of grain to produce about 2 pounds (1 kg) of beef. Most of the grain is turned into manure. The same amount of grain could produce between 40 and 45 loaves of bread.

Grocery shoppers in the developed world have become accustomed to endless supplies of food at low prices. In order to maintain consistent quality and low prices, farmers typically limit themselves to a relatively small selection of cash crops. By limiting farms to certain types of crops, the diversity of the planet is reduced.

Throughout the developed world, food safety is regulated by laws, which protect consumers from dangers such as toxins. Despite the protection that such laws offer to human health, many people worry that industrialized agriculture has resulted in new health risks, environmental damage, and widespread cruelty to animals.

Vegetarianism is gaining popularity in the developed world and is the norm in several cultures. Many vegetarians argue that farm animals should be treated with the same respect as animals kept as pets. Still, most people in the world would prefer to eat meat and eggs and drink milk, but they cannot afford to do so. Is industrial farming the only way to satisfy world food demands? Or is there a "greener" alternative?

▌ **Wheat is the most widely grown cereal grain in the world. It is the staple food for 35 percent of the world's population and provides more calories and protein in the world's diet than any other crop.**

Food Supply and Demand

The world's population has doubled since 1955, and is expected to double again within the next 100 years. Although there has been a huge increase in the number of people in the world, there is still enough food to feed them all. After all, food production has increased over the past 40 years. However, about 1 billion people—or 16 percent of the world's population—live with hunger every day. Poverty, not lack of global food supplies, is the main cause of hunger. Many people in the world simply cannot afford to eat.

The first responsibility of any government is to make sure that its citizens have enough food to eat. Unfortunately, many governments do not meet this responsibility. Money that could be spent on food supplies is often directed toward other pursuits, such as war. Other governments simply do not

■ **Worldwide consumption of red meat grew by 1.57 million tons (1.4 million t) during 2000.**

Many people in the world simply cannot afford to eat.

have the means to produce or obtain adequate food supplies.

A strong farming sector can be the basis for a steady economy. The United States and the **European Union** are major producers and exporters of food. China's agricultural reforms, launched in 1978, paved the way for prosperity in the 1980s and 1990s. Other countries, such as Japan and South Korea, eat well by trading manufactured goods for food.

Opinions differ within the scientific community on whether the world's farmers can continue to grow enough food to meet the demands of the world's growing population. It is not just a question of feeding more people. Diets are changing throughout the world as meat becomes more available and more affordable. Not long ago, meat was a rarity in the Asian diet, which consisted mostly of rice and vegetables. Now Chinese, Malaysian, Korean, and Indonesian consumers eat meat, eggs, and milk, along with their traditional dishes. China's meat consumption is rising by 10 percent each year. Even so, Asia consumes only one-quarter of the amount of meat that the United States does.

Rice is the primary food for half the people in the world. In many regions it is eaten with every meal.

> *Consumers are demanding a less damaging form of food production.*

Experts say that global farm output will have to triple by 2050 in order to keep up with the demand for food. As a result, the urge to spread industrialized agriculture into the **developing world** is growing. At the same time, consumers in rich countries are demanding a less damaging form of food production.

FEEDING THE ANIMALS

The human population has increased rapidly during the last 100 years. Livestock numbers have also risen. In addition to about 6 billion people, the world now has to feed 13 billion chickens, 1.7 billion sheep and goats, 1.3 billion cattle, and 1 billion pigs. More and more of these animals are fed with grains instead of grass, putting extra pressure on global food supplies. About 40 percent of world grain production is fed to animals. China alone has 2.7 billion chickens and more than 400 million pigs, in addition to 1.2 billion people. The economic growth in China has led to a rapid increase in the demand for food—and a change of diet.

Food Supply and Demand

Some scientists warn of a looming food crisis caused by rapid population growth. The crisis will be even greater, they say, when combined with the growing demand for meat in the emerging Asian economies. At the other end of the spectrum are those who argue that by using industrialized agriculture and new technologies, farmers will soon be able to feed about 10 billion people.

The demand for meat has increased greatly with the worldwide popularity of fast-food chains. Meeting such demand, while maintaining standards of quality and taste, can only be achieved through industrialized agriculture. Most large fast-food companies do not purchase beef from small farmers—the costs would be enormous. Small farmers cannot offer the discount prices necessary for cheap food. Big food retailers buy from big producers, who often use whatever means are necessary to offer competitive prices. If genetic engineering can produce better beef more quickly, then so be it.

This logic applies to more than just the fast-food industry. Other suppliers are expected to deliver large amounts of products, too. Often, they must meet strict standards of size, quality, appearance, and price. Whether it is wine or apricots, barley or carrots, there is pressure for consistent quality at the lowest possible price. For the consumer, this results in stores full of a wide variety of foods.

The present global food system favors intensive farming methods. The challenge for agriculture is to produce enough food to meet future demands without damaging the environment or human health.

UNDERNOURISHMENT WORLDWIDE

This chart illustrates the percentage of the population that was undernourished in various subregions between 1996 and 1998. The Food and Agriculture Organization (FAO) estimates that during this period there were 815 million undernourished people in the world, 777 million of which were in the developing world.

Region	Percentage
Middle Africa	50%
Eastern Africa	46%
The Caribbean*	31%
North Africa	29%
South-Central Asia	22%
Western Africa	16%
Southeastern Asia	14%
Eastern Asia	12%
South America	11%
Central America	9%
Southern Africa	7%
Eastern Europe	6%

* The average for the Caribbean is skewed by high malnourishment in Haiti.

KEY CONCEPTS

Population growth The world's population topped 6 billion in 1999, and it increases by about 250,000 each day. People are living longer than they used to. As a result, the world's population is growing at an enormous rate. Experts say that the world can hold about 18 billion people. By 2050, they project that there will be about 9.5 billion people in the world. In order to feed this many people, agricultural practices must change to meet growing demand.

Intensive farming Intensive farming methods allow a relatively small number of people to produce vast quantities of food. Energy, money, and production methods are concentrated on a particular crop or animal, and such specialization generally results in very high output. Technology, in the form of chemicals, genetic engineering, and **hormones**, is used to maximize output. Intensive farming is highly efficient in the short term, producing as much food as possible as cheaply as possible. Intensive farming is the norm in the developed world and increasingly common in the developing world.

Understanding Global Issues—The Future of Farming

An estimated 50 percent of the world's production of fruits and vegetables is wasted.

Factory farming Factory farms are huge operations where animals are kept confined in close-quarters. Some factory farms process millions of animals each year. Feedlots are a typical form of factory farms. Young cattle are brought in and fed grains that promote quick growth. Often, the feed has growth supplements added to further speed-up the growing process. On average, cattle remain in feedlots for 120 days, after which they are sent to slaughter.

Pesticides Pesticides are used to prevent, destroy, or control pests such as mice, insects, fungi, weeds, and bacteria. Many household products, such as rat poison and disinfectants, are considered pesticides, though pesticides are usually associated with the protection of crops. Overuse can lead to resistance on the part of pests, and because most pesticides are chemicals and therefore toxic, they are often dangerous to humans, animals, and the environment. Concern over such risks has led to the development and increased use of biologically-based pesticides, which are generally safer than traditional pesticides, though still not without risk.

Food Supply and Demand

Industrialized Agriculture

Before the arrival of farming, food supply was provided through hunting and gathering. While hunting and gathering could only support small populations, organized farming was able to provide food for large numbers. As farming practices spread, the population grew. Soon, farming was the main occupation of the vast majority of people. Before the Industrial Revolution in the late 1700s, 90 percent of people in Europe worked the land. By World War II, the number of people making their living from agriculture had decreased to about 40 percent. Today, the proportion of farmers in the developed world is about five percent. In the developed world, farming has become a highly industrialized operation, with fewer people becoming farmers each year. The richer the country, the smaller the percentage of farmers. In the United States, agriculture employs less than two percent of the work force.

Mixed farming was once as common in Europe as it is in Asia or Africa. A typical farm produced a wide variety of crops and kept a range of livestock. Mixed farming was an important safeguard against the devastation

Modern farmers tend to be heavily in debt. They borrow large sums to finance the purchase of buildings and machinery.

that could be caused by the failure of a particular crop. Keeping animals was not only useful for milk and meat, but also because animal dung was a valuable fertilizer. Some crops were grown as animal feed for the winter. Horses or oxen were used for working the farm. Farming communities were essentially self-sufficient, producing their own seed, fertilizer, animal feed, and hand tools. Nothing was wasted in the traditional farming system.

In Europe's overseas **colonies**, such as Indonesia, intensive agriculture was already well established by 1900. The plantations that provided sugarcane, tea, coffee, cocoa, cotton, and other farm products were run on industrial lines, alongside the mixed farms of the local population. These crops were grown for export to the colonial powers. Some of these crops have been produced on the same land for centuries.

▬ **Modern farm machinery has played an important role in the industrialization of agriculture. Machines, such as these combine harvesters, are expensive to purchase. The starting price for a combine is $150,000.**

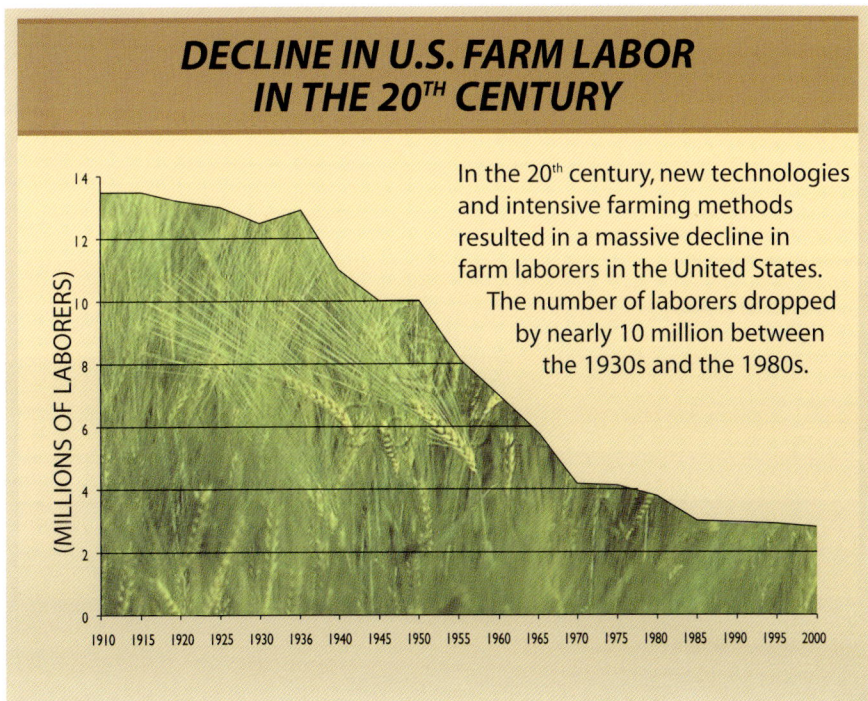

DECLINE IN U.S. FARM LABOR IN THE 20TH CENTURY

In the 20th century, new technologies and intensive farming methods resulted in a massive decline in farm laborers in the United States. The number of laborers dropped by nearly 10 million between the 1930s and the 1980s.

Industrialized Agriculture

Colonial crops were especially vulnerable to pests and disease because they were large plantations of just one crop. Farmers welcomed the arrival of miracle powders and sprays that could kill locusts, tsetse flies, and mosquitoes.

Large-scale cattle farming took hold in many countries in the 19th century and grew rapidly with the invention of refrigerated ships. Farming intensified with the success of mass plantations. In the United States, large-scale farming was a natural result of the Industrial Revolution. Land was plentiful and inexpensive, and over time the open spaces of the U.S. proved to be ideal testing ground for new items such as tractors, combines, grain silos, and crop-dusters.

The two world wars demonstrated the importance of secure food supplies. The wars also strengthened the U.S. position as a food exporter. The arrival of new technologies and improved farming methods pushed the industry forward. Intensive methods could quadruple **yields** while reducing the work force. It was hardly surprising that the developed world embraced the new farming systems with enthusiasm.

People in rich countries eat very well and are able to choose from a vast array of food products. These countries also have the smallest proportion of people working the land. As

> *To damage the soil is to damage the food-bearing capacity of the planet.*

with manufacturing, intensive agriculture requires huge amounts of money. Modern farmers tend to be heavily in debt, since they have to borrow large sums of money to finance the purchase of the buildings and machinery that intensive farming requires. As a result, banks have become part of the driving force behind high production and profit. Farmers who borrow must make interest

Plantation tea pickers can each harvest about 40 pounds (18 kg) of tea leaves a day, enough to make 10 pounds (4.5 kg) of tea.

payments to the banks, so they look to maximize their returns.

One of the risks involved in intensive farming is overuse of land. A downward spiral may be set in motion whereby increasingly intensive methods are needed to squeeze extra production from exhausted soil. Rural unemployment, overuse of pesticides, and contamination of water supplies are among the other risks of intensive farming. There are philosophical and ethical dilemmas, too. Some people wonder whether it is acceptable to treat soil and animals as raw materials for food production in the same way as, say, steel is used for making cars. Animals are conscious beings, and soil is full of **microorganisms**. To damage the soil is to damage the food-bearing capacity of the planet.

KEY CONCEPTS

Industrial Revolution This period, which began in the late 18th century, changed the nature of production by replacing manual labor with machines. Goods that were traditionally created by hand were now being manufactured in vast quantities in factories. The factories promised jobs, and urbanization became a major trend. Farmers began to use methods that increased production and decreased the amount of physical work required.

Genetic engineering Genetic engineering is the process by which the genetic makeup of a particular organism or species is manipulated in order to produce desirable traits or eliminate undesirable ones. Many benefits arise from genetic engineering. For example, plants can be engineered to contain nutrients otherwise unavailable in particular regions, and crops can be modified to be resistant to frost, drought, or certain pests. The full implications of genetically engineered plants have yet to be discovered.

Understanding Global Issues—The Future of Farming

Plantations Plantations are large estates or farms on which a single crop or type of tree is grown. Mainly located in tropical or sub-tropical countries, typical crops produced include coffee, tea, cotton, sugar, and rubber. Labor is usually provided by the local population, who work for low wages, although the plantations are often owned by individuals or companies that live or are based overseas.

Mixed farming Mixed farms produce both crops and livestock. Though production rates are generally lower than on farms that specialize in a particular crop or animal, mixed farming is more environmentally friendly and economical, and farmers tend to be more self-sufficient. Little is wasted—for example, manure is used to fertilize soil, and crops are used to feed livestock. Farmers are protected from the risks associated with monoculture. If poor weather conditions or pests cause particular crops to fail, farmers are still able to earn a living from livestock sales.

Industrialized Agriculture

The industrial approach is best suited to countries, such as the United States, that have fertile soil and reasonably consistent patterns of rainfall. Large fields enable farmers to make good use of giant tractors, combines, aerial spraying, and other industrial methods. Such techniques are much less effective, or cannot be used at all, in arid lands, on steep terraced hillsides, or in forest clearings—environments that make up a large part of the world's farmland.

In the developed world, many farmers are under contract to supply food companies with specified amounts of a certain crop by a certain time. Often, the products delivered must meet specific standards of weight and appearance. Before long, food companies may even demand that farmers use certain seed types, growing methods, and pesticides.

Intensive farming has brought many changes. The most significant change is that now, a relatively small number of large companies provide the seeds, fertilizers, animal feed, machinery, fuel, and drugs on which modern agriculture depends. Some companies are operating in several of these areas, strengthening their influence. Many are investing heavily in the new agricultural revolution—genetic engineering.

A FARMER'S LIFE

The peasant farmer generally has a hard life, characterized by backbreaking manual labor. Most farmers would welcome anything that reduces their workload. Industrialized agriculture has brought about many innovations, including higher-yielding seeds, automated machinery, and pesticides. It would be unreasonable to deny these benefits to farmers in the developing world. The challenge is to use the best new technologies while at the same time avoiding the problems of industrialized agriculture, such as **soil erosion** and over-dependence on artificial seeds, fertilizers, and pesticides. Farmers, once regarded as examples of good health, are now affected by a variety of health problems associated with industrial farming, including stress and pesticide poisoning.

In Japan, rice crops account for 60 percent of pesticide use.

Agronomist

Duties: Responsible for the successful growth of crops by selecting seeds, fertilizers, and harvesting techniques that ensure the best quality

Education: A Bachelor of Science degree with agronomy or plant production as a major area of study

Interests: Science, working with plants, solving problems, and working outdoors

For further information on a career in agronomy head to www.labour.gov.za/docs/mycareer/alphabet/a/agronomist.html or check out agronomy.unl.edu for more information.

Careers in Focus

Throughout their careers, agronomists work in a variety of settings, including outside on the farm collecting soil and plant data, in the office, and in the laboratory analyzing data. These diverse working conditions allow agronomists to effectively study the interaction of plants and soils and find ways to improve the quality and yield of agricultural crops. These scientists help determine the best crops to grow, how to prepare the soil, and how to plant the crops. This field of study, sometimes referred to as crop science, plant science, or soil science, also involves other practices such as irrigating, harvesting, and soil grading. The main goal of agronomy is to find ways to increase the economic output of crops while preserving the environment and conserving natural resources.

Agronomists work closely with others to solve complex issues of pest control, soil quality, and water use. They require excellent interpersonal skills, as they interact daily with co-workers, farmers, and businesspeople. Agronomists often attend conferences and workshops to keep up-to-date on the latest developments in agricultural science.

Not surprisingly, many agronomists are also farmers. Others work as consultants for large agricultural corporations. As a consultant, an agronomist works closely with farm managers, governments, or corporations to develop improved farming methods.

Organizations that employ agronomists include fertilizer and pesticide manufacturers, harvest insurers, university research centers, and various agricultural sectors of the government.

The Global Harvest

About 11 percent of the world's land area is used to grow crops. A further 25 percent is permanent pasture. The world's main crops are grains or cereals, such as wheat, rice, and oats; roots and tubers, such as potatoes and yams; and vegetables, fruits, and beans. Other edible crops include cocoa beans, coffee, sugarcane and sugar beets, and tea. In addition to food crops, there are a wide

The world's main crops are grains or cereals; roots and tubers; and vegetables, fruits, and beans.

range of other agricultural products, including tobacco, cotton, and hemp. In recent years, the demand for oil crops, such as soybeans and sunflower seeds, has increased dramatically in response to the demand for animal feed and for vegetable cooking oil.

Before 1939, thousands of different varieties of crops were grown worldwide. Adapted to local conditions, they were highly resistant to drought and pests. The same applied to fruits and vegetables. Every region had its own varieties. While such variety is pleasing to the palate, it is inconvenient for intensive farming. Large-scale

Fertilizers are used in the Central African Republic to promote the growth of stronger, healthier coffee trees and encourage higher yields.

farm production is only effective when it uses a few of the best varieties. As a result, there has been a decrease in the variety of seeds used to grow crops.

During the 20th century, about 75 percent of the world's variety of crops was lost. The best natural species have been **crossbred** to produce new varieties with certain qualities, such as a high yield or resistance to disease. Farmers around the world have been encouraged to use the new seeds, and most have been happy to do so. The farmer buys **hybrid** seeds from a dealer. The same dealer will often point out that chemical fertilizers will also increase yield and that certain pesticides will reduce damage from disease and pests. Crop production has doubled since 1965, but use of fertilizers

> *During the 20th century, about 75 percent of the world's variety of crops was lost.*

and pesticides has increased even more.

Before long, the farmer, who was once self-sufficient becomes locked into a system that depends on outside supplies of seeds, fertilizers, and pesticides. If yield goes down, the solution is to apply more fertilizer. If pests eat the crop, the solution is to spray more often. This system of farming is extremely convenient for farmers, since they do not have to spend time hoeing weeds and collecting seeds. It also produces higher yields, for hybrids have larger grain size and there is less damage from disease or insect infestation.

BIOCIDES

Plants can be damaged by a wide variety of parasites, diseases, and pests. As a result, most farmers welcomed biocides, which are chemicals used to kill living organisms. A field of golden wheat without weeds is a sure sign that biocides are being applied. Early pesticides were crude and sometimes dangerous. A breakthrough came with the development of chemicals such as DDT. Although it is now banned in the developed world, DDT was once considered an important pesticide. It was particularly good at killing mosquitoes. Today, we know that DDT remains in the soil and is toxic to animals, not just mosquitoes. Modern pesticides are designed to be **biodegradable** and harmless to humans. However, there is concern that the widespread use of some chemicals in food production could be affecting human health. One of the most worrying possibilities is that certain chemicals can imitate the hormones of living creatures, affecting how they reproduce.

Developed countries have strict regulations covering the use of biocide chemicals, but the large variety of ways in which biocides are used makes it impossible to rule out health risks. The question is how to balance the risks against the benefits. Some of the most widely used early pesticides are now banned in many developed countries. There is now recognition that pesticide use needs to be minimized and that it can be self-defeating, since each chemical weapon soon produces a counter-weapon from nature as organisms become resistant. Even so, the world market for pesticides is currently about $30 billion a year and growing.

The Global Harvest

CROP CONCENTRATION

The grain business has become highly concentrated, with the United States being the main player. The U.S. is the largest producer of hybrid seeds, the greatest exporter of grain, and home to the biggest agribusiness companies. Corn is native to Mexico, but 36 percent of world production takes place in the U.S. At the same time, American grain production relies on a small number of grain types. Similar concentration of crops is taking place elsewhere. Southeast Asia once had about 100,000 varieties of rice. A single hybrid variety is now grown in 65 percent of paddy fields. Modern hybrids account for half of all the corn, wheat, and rice grown in Asia. Production has rocketed, but so has dependence on external inputs, such as fertilizers and pesticides.

Nearly 40 percent of the world's grain is fed to livestock.

The battle against crop disease is far from over. Between 20 and 30 percent of all harvested food crops are still spoiled by disease or pests. An even higher percentage of crops are destroyed before they can be harvested. New chemicals that help reduce crop losses are good news for farmers.

While the evidence that pesticides have damaged human health in the developed world is controversial, there is no doubt that, in the developing world, many deaths and illnesses are the result of the misuse of farm chemicals. Farmers in the developing world may not be able to read the instructions and think that the more they apply, the more effective the chemical will be.

Food sprayed with chemicals may also be imported by developed countries. The U.S. government found that illegal residues on imported foods were four times as high as those found on domestic foods.

KEY CONCEPTS

Fertilizers Farmers have always used fertilizers to enrich the soil and improve plant growth. Before the Industrial Revolution, all farming was organic and depended on the recycling principle. Animal manure and dead plants were plowed back into the soil, adding nutrients while getting rid of unwanted waste. Sometimes artificial fertilizers were used, but they were made from potash and guano (bird-droppings), not chemicals. Though **nitrogen** occurs naturally, artificially produced nitrogen compounds, such as nitrates and urea, have proved to be particularly effective fertilizers, and their discovery transformed agriculture. It meant that crop production could be increased with the use of chemicals. Modern fertilizers provide the soil with nitrogen, potassium, and phosphorus. Though these elements are now considered essential, soil fertility depends on many other factors, such as the number of microorganisms and worms.

Understanding Global Issues—The Future of Farming

Organic Farmer

Duties: Produces food without the use of synthetic fertilizers, pesticides, and drugs
Education: A graduate degree in organic horticulture or agriculture
Interests: Human health, the environment, and food production

Navigate to the Organic Farming Research Foundation's Web site: www.ofrf.org/index.html or check out an environmental horticulture program at hort.ifas.ufl.edu for more information.

Careers in Focus

The decision to become an organic farmer is often as much a lifestyle choice as it is a career choice. Organic farming requires a strong commitment to producing agricultural crops that are socially, environmentally, and economically sustainable. Organic farmers must farm without the use of toxic chemicals, preservatives, or **irradiation**.

Organic food producers must comply with strict federal standards, which are verified yearly by private or governmental authorities. To maintain certification, farmers must practice long-term soil management, prevent pesticide contamination from near-by farms, and keep detailed farm records. Farmers must also meet standards for cleanliness, storage, and pest control as stated in the Organic Food Production Act, which was established in 1990 by the United States Department of Agriculture (USDA).

Organic farms are often similar to traditional family run farms. While organic farming uses traditional knowledge, modern farmers require knowledge of the latest organic farming methods. Most organic farmers possess a graduate degree in organic horticulture or agriculture. They usually own their own farms or work for large organic farms.

In recent decades organic farming has become recognized as a viable alternative to intensive farming practices. Public demand in Europe and North America for organic foods increases every year.

Livestock Farming

The eating of one species by another is an essential part of the web of life. Nobody expects a lion to be vegetarian. Early humans learned to be omnivorous as part of the battle for survival. As hunters and gatherers, humans needed a large territory to support a family. With the arrival of farming there was a rise in the domestication of animals. Instead of hunting animals in the wild, farmers now bred animals in captivity.

Meat was the food of choice in most human societies. For a hunter, the search for meat could be long, difficult, and often unsuccessful. As a result, hunters learned to respect their prey. For the family that raised and slaughtered its own animals, there was at least an awareness of the animal's condition, even if there was no regard for animal rights. Grocery shoppers in the developed world have become totally removed from the process

Some feedlots in North America contain more than 50,000 animals.

that turns animals into meat.

The industrialization of agriculture has been hard on livestock. Industrialization resulted in highly efficient systems for producing vast amounts of inexpensive meat. Cattle used to be placed in pastures and left until they reached a certain weight. Now they are fed quick-growth hormones and concentrated feed. Instead of allowing animals to run around in fields, where they build up muscles, their flesh is kept soft by confining

CHICKEN FARMING

The chicken is perhaps the most abused animal on Earth. Modern factory farm egg production involves keeping thousands of birds in small cages, where they spend their lives perched on sloping wire mesh under artificial lighting. They lay about 300 eggs in one year—5 to 10 times as many as an average free-range chicken. After about a year, when their egg-laying rate becomes uneconomic, they are slaughtered and ground up for soup, pies, or animal feed.

Broiler chickens are different from chickens that lay eggs. They reach slaughter weight in 40 to 45 days—twice the growth rate of their ancestors. Kept in large sheds containing 20,000 birds or more, they are collected for slaughter by handlers who pick up several at a time by their legs. Because of their lack of exercise, their weak leg bones sometimes break. The chickens are carried off to the processing plant, where they are shackled by the legs to a conveyor. On the conveyor, they are rendered unconscious in an electrically-charged water bath, have their throats cut, and are then lowered into scalding water to loosen their feathers. The Humane Slaughter Act, which was created to protect animals from unnecessary pain, does not apply to chickens.

▬ **About 95 percent of the eggs produced in the United States come from "egg factories."**

Livestock Farming

▮ Worldwide, people eat more than 73 billion pounds (33 billion kg) of chicken meat annually.

them to narrow stalls. The meat of a veal calf is more highly priced if it is pale in color. A liquid, iron-free diet will keep it that way, but the growing calf will not receive essential nutrients.

Intensive farming has forced animals to produce more than their bodies were designed to produce. A typical dairy cow in the western world produces 10 times as much milk as it would if it were left alone in a pasture.

In confined dairy farms, the strain on the cow is enormous. Dairy cows used to live an average of 15 years or more. Today, dairy cows are worn out after only six or seven years. They are then killed and ground

BUSH MEAT

In Africa, the forest is often called the bush, and meat from the forest is known as bush meat. Bush meat can come from a variety of animals, including monkeys, chimpanzees, gorillas, elephants, and leopards. Roads built by logging companies make it easier for hunters and poachers to move into previously untouched wilderness areas. With improved access to wildlife, the sale of bush meat has increased dramatically. The meat is sold in markets around the world as an exotic delicacy. It is also popular in poor communities where people cannot afford other foods. In some countries in Africa, bush meat is up to 75 percent cheaper than farmed meat, such as beef. This commercial hunting of wild animals is one of the most serious threats to the survival of African wildlife, and has already resulted in local extinctions of several species.

FROM GRAIN TO MEAT

Grazing animals convert grass into edible meat and milk for humans. Pigs and poultry were traditionally fed human food waste. Today, intensive farming uses grain, oilseed, and fishmeal products to feed livestock. It takes four to seven pounds (1.8–3.2 kg) of grain to produce one pound (0.5 kg) of pork.

The pharmaceutical industry is heavily involved in the livestock industry. In the United States, livestock consume about 40 percent of all the antibiotics used in the country. Most of these antibiotics are administered in feed.

About one billion pigs live on farms throughout the world. Approximately half of these are raised in China, the world's leading pig-producing country.

up for use as hamburgers or "animal protein." Until the outbreak of mad cow disease—a brain disease affecting cows—some of this material, along with slaughterhouse waste from pigs, sheep, and chickens, was even fed back to cattle—though most farmers did not know that some of the "protein" in their feed came from cows.

In terms of the numbers involved, pigs and chickens have suffered the most from the industrialization of farming. So far, sheep have largely escaped the factory farm approach, but genetic engineering may change all that.

In terms of the numbers involved, pigs and chickens have suffered the most.

Farm animals are routinely given antibiotics to accelerate growth and combat infection. The fear is that humans who eat products made from the flesh or milk of these animals are also ingesting low dosages of antibiotics. Overuse of livestock drugs may be contributing to the rise of drug-resistant bacteria.

In theory, a factory farm animal raised in hygienic conditions should be less vulnerable to many of the insects, parasites, and diseases that affect animals living in the open countryside. In reality, most factory farm animals live with stress and disease. **Salmonella**, common in chickens, can be passed on to humans. Ironically, the increase in poultry consumption in the developed world is partly due to health concerns. Lean, white chicken meat is thought to be healthier than red meat. In poorer parts of the world, chicken and pork are popular because they offer a more efficient way of converting grain into meat. They also provide a cheaper form of meat than beef or mutton.

Intensive production of poultry and pork has made meat affordable to millions of people who previously had vegetarian diets—not from choice but from poverty. Meanwhile, vegetarianism is gaining ground in the developed world, partly for health reasons, partly as a matter of conscience.

Another trend in livestock production is toward greater variety in meat consumption. Turkeys and ducks have been farmed intensively for some years. Ostriches have recently joined the ranks of livestock

> *Chicken and pork are popular because they provide a cheaper form of meat than beef or mutton.*

raised for meat. Game has long been on the menu, ranging from grouse in Great Britain to gorilla in Cameroon. The notion that meat comes from a handful of familiar farm animals, such as cattle, sheep, and pigs, may be about to change. From kangaroos to zebras, there are plenty of other animals to choose from.

Wild animals show clear signs of stress if confined. Even long-domesticated animals keep habits from the wild and suffer from close captivity. Stalls so narrow that an animal cannot turn around, a common find on pig farms, are clearly stressful and could affect meat quality. At the same time, intensive feedlots produce huge amounts of **excreta**. Pigs in the United States produce 20 times as much excreta as the human population and are not even connected to sewage systems. The problems presented by animal waste are not unique to the United States. Ammonia from the Dutch livestock industry is thought to be responsible for 30 percent of that country's acid rain.

KEY CONCEPTS

Animal rights Animal rights activists believe that, like humans, animals have the right to be treated with respect and without cruelty or exploitation. People who support animal rights oppose factory farming, where animals are treated as products rather than creatures that require appropriate living conditions.

Broiler chickens Broiler chickens are raised to produce meat rather than eggs. There is an enormous worldwide demand for chicken meat—approximately 30 billion broiler chickens are produced yearly. Broiler chickens are often reared in large, overcrowded sheds and many do not make it to slaughter, succumbing to disease, injury, and even cannibalism. There are many food safety risks associated with broiler chickens raised in intensive confinement, and there are a number of associated environmental concerns as well. High concentrations of dead birds and manure result in toxic soil and water, making the industry a leading polluter.

Antibiotics Antibiotics fight infectious disease by destroying disease-causing bacteria, with little or no harm to the host. Though used to successfully treat once-fatal diseases, such as scarlet fever, antibiotics are becoming increasingly ineffective. Used to stimulate growth and control disease in livestock, the widespread use of antibiotics in agriculture has resulted in increased drug resistance, reducing the effect of antibiotics in the fight against disease. This has consequences for humans as well as animals. Resistance genes can not only be transferred between animals, they can also be transferred from animals to humans, through the consumption of meat and poultry.

Understanding Global Issues—The Future of Farming

Rancher

Duties: The care and production of livestock

Education: While ranchers often learn on the job, many universities and colleges offer bachelor's degrees in farm management or applied agricultural technology and entrepreneurship

Interests: Entrepreneurial spirit and a passion for raising animals

Navigate to the Ranch Work's Web site: www.ranchwork.com, or check out www.beef.org for more information on ranching.

Careers in Focus

A rancher is a farmer who raises cattle, horses, and other livestock on large areas of grazing land. One of the most important roles of the rancher is to monitor the quality of range grasses consumed by livestock. Careful recording of grass quality ensures the well-being of livestock and prevents the overgrazing of valuable ranch grasslands. Another important aspect of ranching is caring for the health of livestock herds. Ranchers routinely vaccinate their animals against disease and provide medical attention when they are ill. Every fall, ranchers are also involved in branding their livestock. A mark is burned onto the animal's hides to identify them as the property of the rancher. This mark, or brand, also helps ranchers to monitor the age and health of animals.

Modern ranchers often employ the latest technology. Some ranchers round-up livestock into corrals with the aid of helicopters. They also use computers to obtain cattle prices and manage ranching operations. Today's ranches are typically large commercial operations that require massive monetary investments in land and machinery. As a result, ranchers often possess excellent business management skills. In addition to raising livestock, ranchers are increasingly involved in the development of sophisticated meat packing technology, and the export of meat products throughout the world.

While most ranchers raise traditional stock such as cattle, sheep, or horses, some ranchers also breed other animals such as ostriches, bison, or elk. Many ranchers are also becoming interested in sustainable ranching. One aspect of sustainable ranching focuses on the improvement of grasslands used for grazing, which works to increase animal health and nutrition, as well as maintain the ecological balance of the land.

Map of Cereal Production

Figure 1: Total Cereal Production by Country in 1999
(million tons)

- ☐ No data
- under 10
- 11 - 22
- 22 - 33
- 33 - 44
- 44 - 55
- 55 - 66
- over 66

Low cereal production can give a misleading impression of the productivity of farming in developing countries. Many different crops can be harvested, including local varieties of fruits, vegetables, and nuts. In sub-Saharan Africa, for example, traditional farmers grow more than 2,000 different food plants.

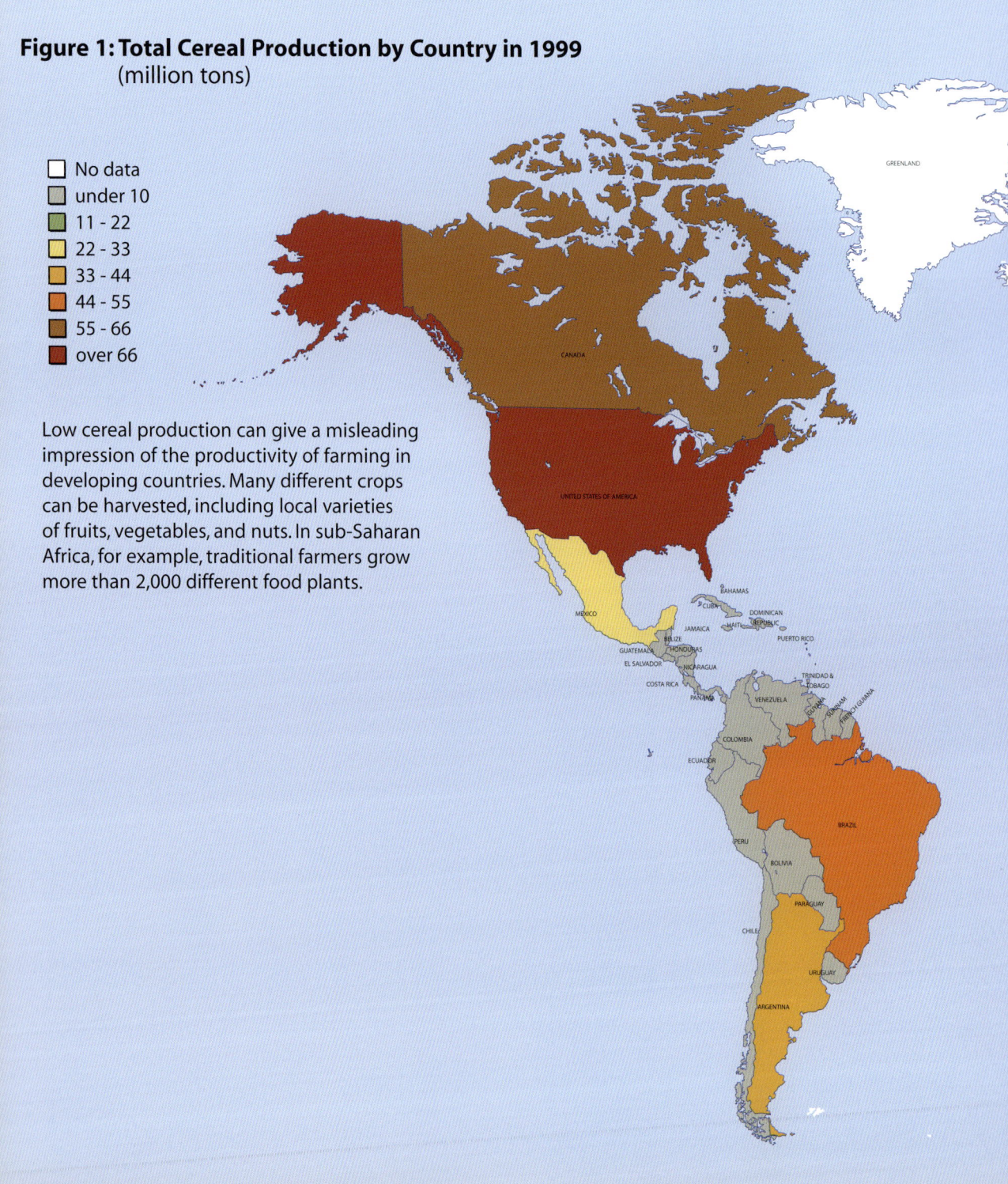

Understanding Global Issues—The Future of Farming

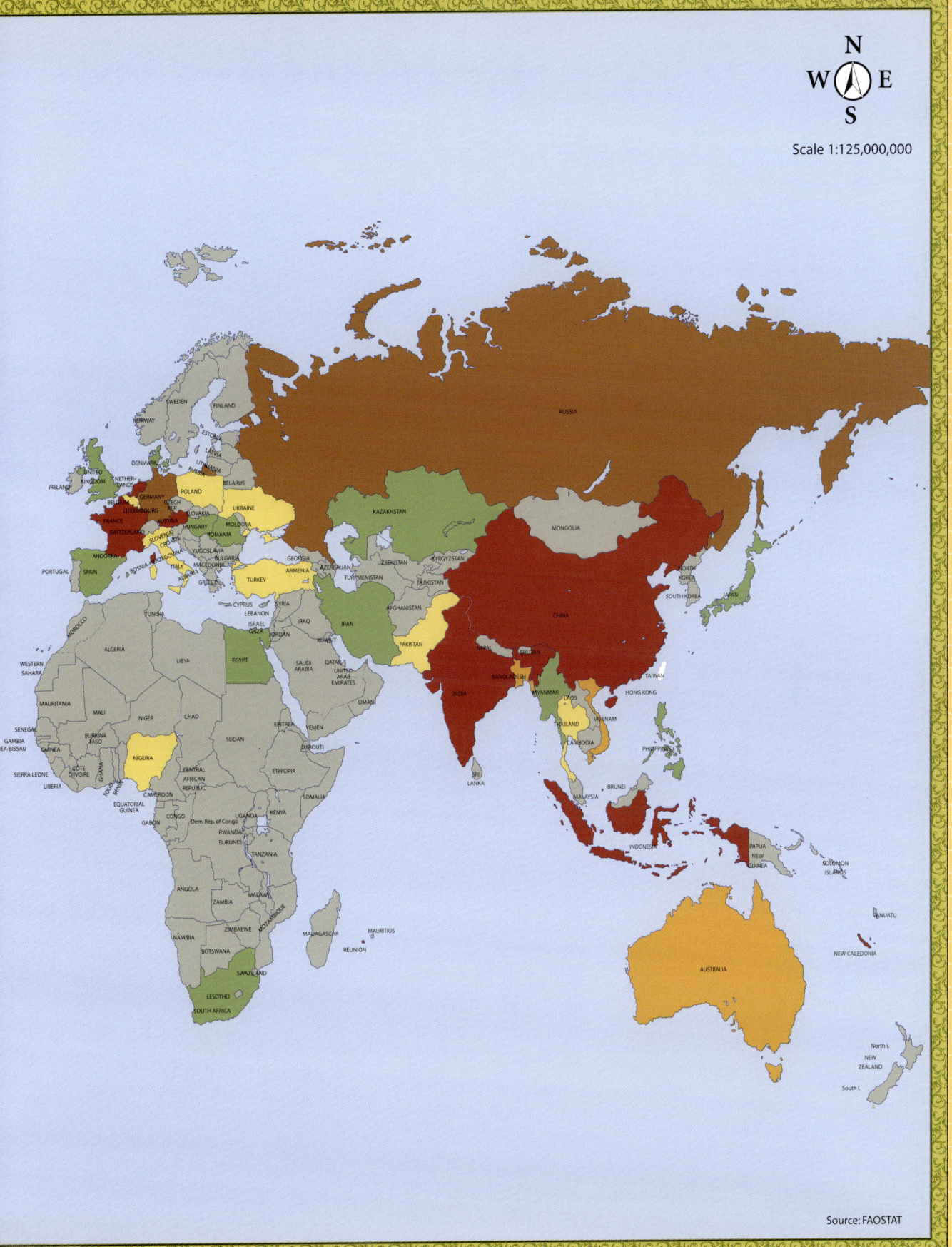
Map of Cereal Production

Charting the World's Food Production

Figure 2: Global Index of Food per Capita (1963-1997)

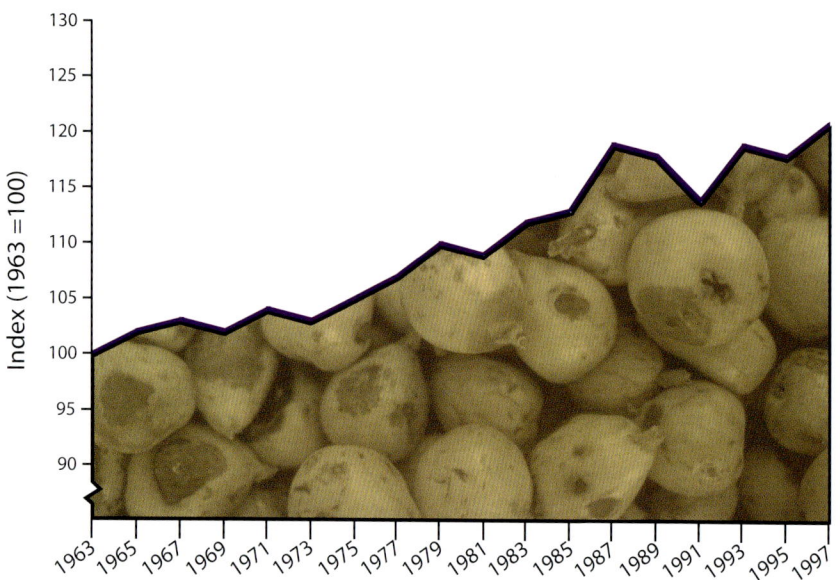

Global food production has increased nearly every year since 1965. However, the production of food per capita has been more erratic as a result of population growth. Not all regions of the world have seen increases, food production per head in sub-Saharan Africa has fallen, while in Asia it has risen dramatically.

Figure 3: Global Soil Quality
(percentage of total world land area)

Only 11 percent of the world's soils are suitable for farming without irrigation, drainage, or other improvements.

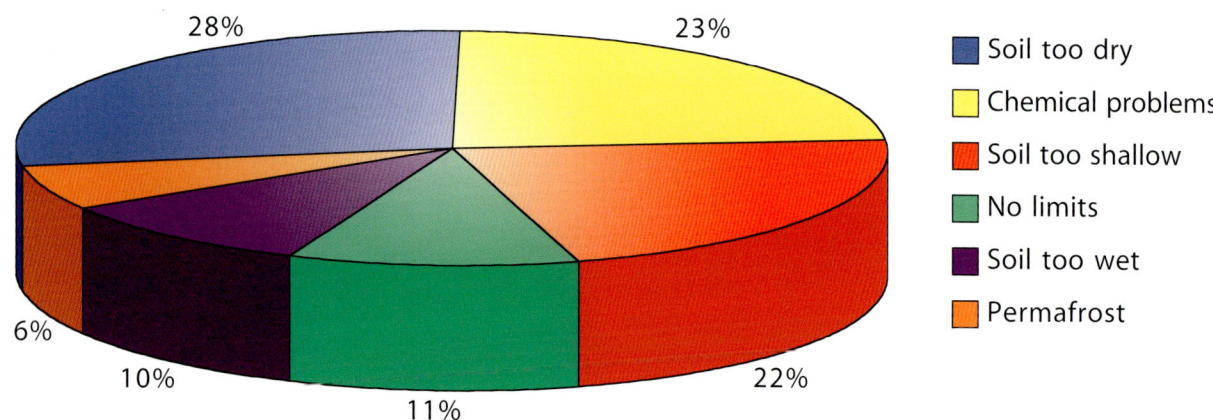

- Soil too dry — 28%
- Chemical problems — 23%
- Soil too shallow — 22%
- No limits — 11%
- Soil too wet — 10%
- Permafrost — 6%

Understanding Global Issues—The Future of Farming

Figure 4: World Pesticide Consumption (1983–1998)

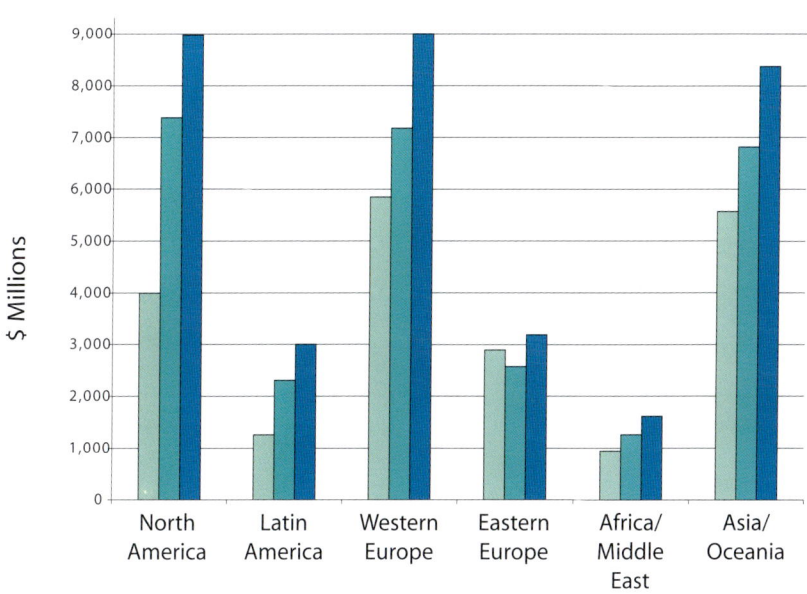

The global use of pesticides continues to rise at an alarming rate. In North America alone, the amount spent on pesticides doubled between 1983 and 1998.

■ 1983
■ 1993
■ 1998

Figure 5: Leading Producers of Cereals in 1999
(million tons)

Three countries dominate world cereal production—China, the U.S., and India. China is by far the biggest producer of cereals, providing for its population of 1.2 billion people. India, with some 1 billion people, also has to grow enormous amounts of food. The U.S., by contrast, produces far more food than its population can eat.

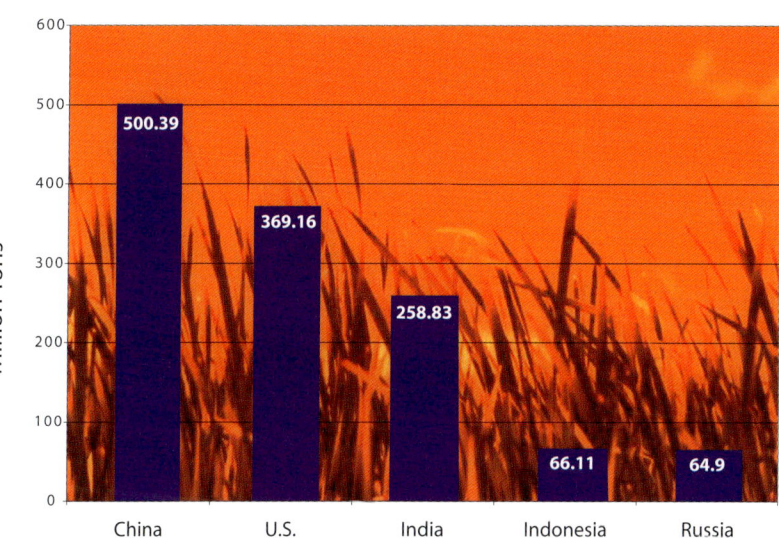

Charting the World's Food Production

Farm Trade and Subsidies

The international grain trade was well established by the time of the Roman Empire. International trade also took place in dried fruit, salted fish and meat, spices, and wine. Cocoa, coffee, and tea were added to the list of international trade commodities after they were brought back from the New World by the European voyages of discovery. In the 20th century, methods of processing and refrigeration enabled perishable foods to be transported long distances and traded worldwide. In the United States, food now travels an average of 1,240 miles (2,000 km) from farm to plate.

Grain is the foundation of the food supply system. While

> *In the food market, taste has taken second place to appearance.*

richer people demand more variety and taste, peasant food was, and still is, largely grain-based. Before the Industrial Revolution, most food trade was local. Produce was grown and eaten within the same rural communities or transported short distances to be sold in the marketplaces of local towns and villages. Longer-distance trade was restricted to products such as dried wheat, spices, and wine. Trade in salt, sugar, coffee, and cocoa helped to build the wealth of western Europe. When canning and refrigerated ships became available, the trade in food rapidly expanded. Today, even fresh fruits and vegetables are routinely shipped around the globe.

International trade in farm produce is now very big business. The trade is dominated by a relatively small number of large companies, or multinationals. Globalization is further concentrating the power of the top 20 companies. These companies control much of what is grown and eaten in the developed world. Their influence is now spreading to developing countries as well, helped along by free trade. The leading food corporations have interests not only in farm produce but also in food processing, fertilizers, pesticides, and the new

New packaging technologies, such as breathable bags, have dramatically increased the shelf life of fresh produce.

Wooden grain elevators are used to store grain before it is loaded into rail cars for shipping. Today, many of these aging elevators are being replaced with massive "super elevators."

technologies associated with genetic engineering.

In the past, most food was locally produced by millions of farmers using traditional methods of cultivation. Apples or potatoes grown in one district were different from those grown in another. Such diversity does not suit the modern market. Retailers and consumers expect consistency and quality of appearance, as in round green apples and evenly shaped potatoes. Most consumers will not buy farm produce that does not meet these standards. Taste has taken second place to appearance. Those in favor of genetic engineering claim that genetic manipulation may be able to produce fruits and vegetables that not only look good but have more flavor too. The U.S. market has already seen the launch of such a tomato—the Flavr Savr, genetically altered to delay ripening so that it

FARM SUPPORT

The European Union spends $50 billion a year on farm support. Only two percent of this is used to encourage environmentally friendly farming. The rest is spent on promoting intensive agriculture. Government policy in many other parts of the world has encouraged farmers to use intensive methods, too. In the developed world, **subsidies** have also resulted in huge stockpiles of food supplies as many governments pay to keep farmers in jobs, even when there is no demand for the products. To counteract food surpluses in Europe, the European Union has begun paying farmers to not produce crops on certain areas of land. This land is known as "set-aside."

Farm Trade and Subsidies

spends longer on the stem and develops more taste.

In the interests of national food security, many governments subsidize farmers. Subsidies often include payments based on production, low-cost loans, tax reductions, and inexpensive supplies of fertilizer and pesticide. The sum total of these subsidies is enormous. Most of these benefits have gone to those running large farms. Subsidies have contributed to the industrialization of agriculture by promoting intensive methods of farming. The more farms produce, the more subsidies they receive. Intensive methods are usually designed to increase production, with little regard for environmental effects or animal welfare. Only in recent years have some subsidies been used to encourage environmentally friendly farming. Subsidies have remained despite free

> *Only in recent years have subsidies been used to encourage environmentally friendly farming.*

market pressure to get rid of them. While farm subsidies are gradually being reduced all over the world, New Zealand is one of only a few countries that has removed farm subsidies altogether.

Health and safety measures are also being streamlined to provide globally acceptable standards. The new rules are supposed to benefit poor farmers, but the main beneficiaries are likely to be the multinationals, who want as free a trading environment as possible. Since they can switch headquarters from country to country, they could, in theory, move to the location with the fewest regulations. Though farming operations themselves can only be moved to countries with comparable soil quality and

TOP 10 SEED COMPANIES

	COMPANY	COUNTRY	SEED SALES ($ MILLIONS)
1.	DuPont (Pioneer)	U.S.	1,938
2.	Pharmacia (Monsanto)	U.S.	1,600
3.	Syngenta	Switzerland	958
4.	Groupe Limagrain	France	622
5.	Grupo Pulsar (Seminis)	Mexico	474
6.	Advanta (AstraZeneca and Cosun)	UK and the Netherlands	373
7.	Dow (and Cargill North America)	U.S.	350
8.	KWS AG	Germany	332
9.	Delta and Pine Land	U.S.	301
10.	Aventis	France	267

Food trade is a major global business, dominated by about 20 companies. Many small farmers have been forced out of business by competition. In the United States alone, more than four million small farms have gone out of business since the 1930s. Increasingly, fewer companies control more of the market. In 2000, the top 10 seed firms controlled 30 percent of the world seed market.

climate, there is plenty of room for food-purchasing companies to switch sources of supply. For example, farmers in California were hard hit by Del Monte's decision to move their peach-growing operations to Italy and South Africa. The globalization of agriculture and food trade is making such decisions more common. To stay in business, farmers have to compete not only with local farmers but also with overseas farmers paying or working for lower wages and enjoying climatic advantages. In these cases, intensive farming methods may be the only chance farmers have of competing on a global scale.

Indonesian farmers have increased rice production in their country by planting high-yielding varieties of rice. The effort was initiated by the government in the 1960s.

KEY CONCEPTS

Globalization During the latter half of the 20th century, new information and communication technologies, such as the Internet, combined with new trends in international commerce to produce an interconnected global marketplace.

Multinationals Multinationals are large companies with operations in more than one country. Multinationals are able to move their operations to countries where production and labor is cheaper, or, in the case of agriculture, where the climate and other conditions allow for higher yields and cheaper production. As a result, many farmers, particularly those in the developed world, are under pressure to turn to intensive farming in order to compete. Farmers in the developing world are increasingly turning away from traditional farming methods in order to incorporate methods favorable to multinational corporations.

Farm Trade and Subsidies

Who's Afraid of Biotech?

Crossbreeding of plants and animals has been carried out for thousands of years. For example, from the wolf humans have bred an astonishingly diverse range of dogs. Cattle, horses, pigs, sheep, goats, and poultry have all been bred for qualities such as strength, size, and vigor. After World War II, crossbreeding of grain plants produced high-yielding varieties. The discovery of DNA and the unlocking of the genetic code have taken breeding possibilities into a new realm.

Most people find genetic manipulation more acceptable when applied to plants than to animals. The idea of a mouse with an ear growing on its back or a pig bred to provide human transplant organs repulses many people, regardless of the scientific benefits. There is much debate about the ethics of biotechnology. While many people wish it had never been developed, there is little chance of turning back the clock.

For farming and food production, the potential of biotechnology seems limitless, so it is not surprising that leading food and agricultural companies are eager to be part of this latest revolution. Although biotechnology is a current issue, it is not entirely new. Producers of bread, cheese, and beer have

The idea of a pig bred to provide human transplant organs repulses many people, regardless of the scientific benefits.

been using microorganisms, such as bacteria, to produce food for thousands of years. There are many other potential benefits. Plants can be engineered to resist disease or pests, have higher nutritional content, and a longer shelf life. Other projects in the experimental stages include frost- and drought-resistant plants.

Today, the main effort is directed at developing herbicide-resistant plants. Australian scientists were among the first to develop transgenic tobacco and cotton. These plants are resistant to a common herbicide called 2-4D. This biodegradable chemical is often sprayed on grain crops and can damage other plants growing nearby. Having crops that are resistant

Genetically modified fruits and vegetables have a uniform appearance that makes them appealing to consumers.

A PATENT FOR NATURE?

In 1997, the U.S. Patent Office awarded a **patent** to the Texas firm RiceTec Inc. for the strains of rice it had developed by crossing American dwarf varieties of rice with basmati rice. RiceTec marketed this new strain as "American basmati," or "basmati style." RiceTec claimed that their strain could be considered an invention and therefore deserved a patent. Opponents to patent rights on crops and plants, such as ActionAid, claimed that RiceTec was simply attempting to claim ownership over nature. Basmati rice had previously been developed over centuries by farmers in the Punjab region of India and Pakistan. The patent attracted widespread criticism and was considered a prime example of the dangers of biopiracy. Biopiracy is the commercial exploitation by companies in the developed world of the traditional knowledge and resources of the developing world. The Indian government, along with ActionAid, challenged the patent. As a result, the U.S. Patent Office revoked the vast majority of RiceTec's claims to basmati rice. However, RiceTec can still grow three varieties of basmati in the United States.

to herbicides means that weeds can be sprayed without damaging crops.

Another use for biotechnology is to produce synthetic substitutes for items such as vanilla, cocoa, and coffee. However, synthetic substitutes would seriously damage the economies of some developing countries. When synthetic rubber was invented, the economies of Brazil and Malaysia suffered. Rubber occurs naturally in these countries, and many people were once employed in the rubber industries there. Now, rubber can be produced locally and less expensively.

Sugar substitutes are yet another candidate for synthetic production. Syrup from corn has already replaced sugarcane fructose throughout much of the U.S. market.

Plant-based products can also be engineered to have the taste and consistency of chicken, steak, or pork. There may be some barriers to this, including price, consumer interest, and industry acceptance. The livestock industry, and farming in general, would be greatly affected by the production of meat alternatives. Perhaps a more likely trend is toward new crops grown for use as fuels and plastics. New transgenic crops on the market include virus-resistant squash and potatoes with a built-in insecticide.

Biotechnology companies must often compete for patents.

Farmers must decide between planting genetically modified crops or traditional crops.

KEY CONCEPTS

Biodiversity Biodiversity refers to the variety of plant, animal, fungus, and microbial species found on Earth and the various ecosystems, or habitats, in which they live. Evolving over time, organisms have become uniquely adapted to particular ecosystems, and the removal or extinction of any one species can have far-reaching and often destructive effects. If an ecosystem's delicate balance is upset, air quality, soil quality, and even rainfall can be affected. As habitats are destroyed, hundreds of thousands of species—and potential medical cures and nutrient-rich foods— are forever lost.

Genes Genetics go a long way in determining how living things look, behave, and even what diseases they are prone to. Found in the cells of every living organism, genes provide a blueprint for the development of an individual organism. Genes are coded messages carried within deoxyribonucleic acid (DNA),

THE GM DEBATE

One of the most controversial issues facing the agricultural industry in the 21st century is the debate over the benefits and dangers of genetically modified food. Many scientists and large biotechnology companies promise a GM revolution that will drastically reduce world hunger, decrease pollution, and even help cure disease. On the other side of the argument, concerned citizens, scientists, and environmentalists are warning of a crisis if GM crops replace traditional varieties.

GM food will help feed the world's hungry by increasing the yield of crops.	There is already enough food to feed all the people on Earth. In fact, some areas of the world have massive surpluses. It is poverty, not lack of supply, that causes hunger.
Herbicide and pesticide resistant crops will make farming easier by allowing farmers to spray large areas without damaging crops.	Herbicide resistant crops could cross-pollinate with weeds, creating "superweeds."
Crops can be designed that are poisonous to pests, reducing the farmer's reliance on pesticides.	Poisonous crops could also kill non-pests.
Crops can be designed that may help cure disease.	Introducing new genes into food may cause unforeseen allergies and other side effects.

a molecule found in all cells, and are passed down from generation to generation. Sometimes genes mutate, and, if passed on to the next generation, they can influence the traits and development of individual organisms and, eventually, the entire species.

Transgenic An animal, plant, or even a microorganism that contains the genes of a foreign species is described as being transgenic. Genetic engineering allows desirable genes to be artificially transferred from one organism or species to another, resulting in, for example, disease-resistant crops, longer-lasting fruit, and even animal organs that can be transplanted to human beings. Transgenic plants are often referred to as genetically modified (GM), and are sometimes even called "Frankenfoods."

Patents are awarded for developments such as transgenic plants and organisms. In other words, those who invest in research want some protection for their inventions. Critics argue that patenting life forms is unacceptable. Despite this, some patents have already been granted. In 1992, a company called Agracetus received patents for transgenic cotton. Many more patents for genetically modified plants have been granted since.

Genetic engineering of animals is even more controversial. The featherless chicken and the self-shearing sheep are among the more bizarre projects being worked on. Other research is directed at the use of farm animals as pharmaceutical factories. These animals produce medically useful substances, such as **insulin** or **hemoglobin**. GM vaccines are being developed to protect livestock from disease. Genetic manipulation can make pigs less prone to stress. This would help them cope with the pressures of intensive farming methods.

Private companies conduct most of the biotech research carried out on food. Naturally, they focus on projects that are likely to make them the most money. There is less incentive for a company to develop crops or livestock that are resistant to disease, since it is often the same companies that manufacture biocides.

Some countries have introduced grain-testing programs to detect the presence of genetically modified grains.

There are many risks involved in releasing new plants that are resistant to herbicides or insects. It is possible that these invincible supercrops could pollinate wild plants to create new herbicide-resistant weeds. Plants with built-in toxins that protect them from pests might poison other living things, such as birds, mammals, or bees. Widespread use could possibly produce toxin-resistant pests. Using genetically engineered plants in agriculture could release new genes into the environment. Wildlife might eat the new plants, which could alter the food chain. The effects of these changes are largely unknown.

Understanding Global Issues—The Future of Farming

Plant Scientist

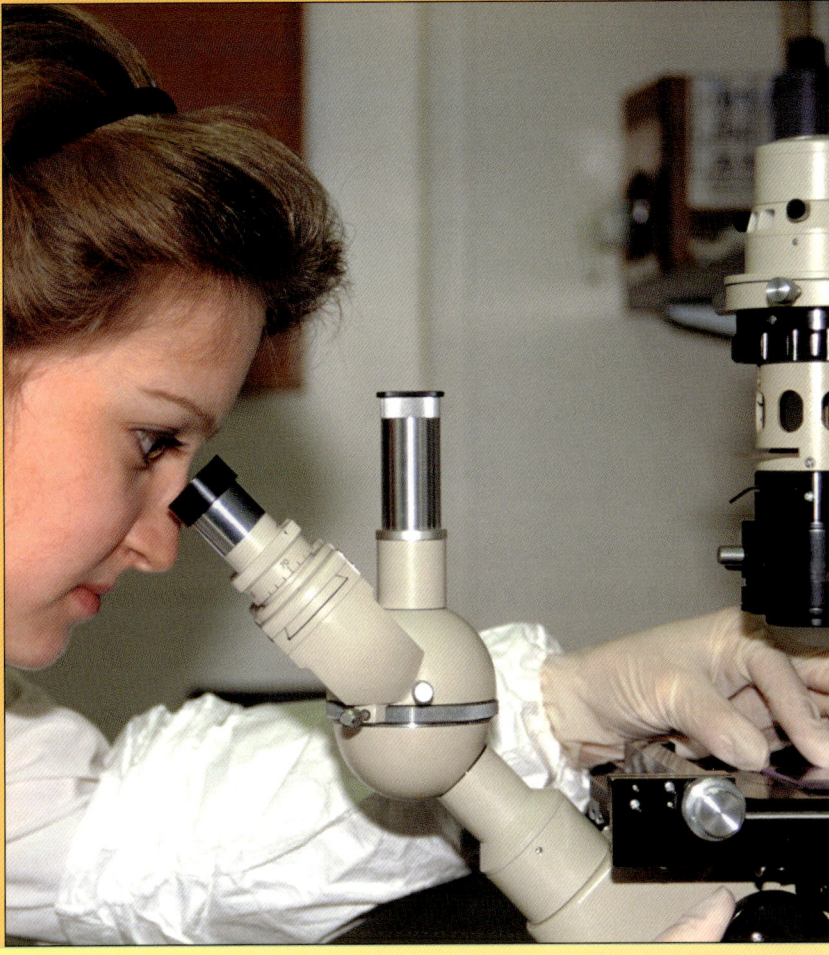

Duties: Develops new or improved varieties of plants through the use of biotechnology
Education: A master's degree in biological science or botany
Interests: Passion for working with plants and biotechnology

Navigate to a plant science career at: www.nal.usda.gov/bic/Education_res/iastate.info/bio2.html or check out the FAO employment Web site at: www.fao.org/VA/Employ.htm for more information.

Careers in Focus

Plant scientists working with biotechnology develop new plant strains by altering a plant's genetic material. This specialized field of genetic engineering involves a complex process of extracting the genetic material from healthy plant cells to create new plants with improved traits. Many scientists hope that the products engineered from these new technological practices will help increase food productivity and food quality in countries around the globe.

One of the most exciting areas of plant science research is the combination of biological technology with agricultural practices to improve food quality and consistency. A major goal of plant science is to produce plant strains that are healthier and have increased resistance to pests and disease.

Plant scientists may be self-employed, or work as researchers in universities, large biotechnology corporations, or non-profit organizations, such as the United Nations' Food and Agriculture Organization (FAO). The FAO has become a global leader in the rational and scientific investigation of GM foods. Plant scientists with the FAO work to ensure that farmers in developing nations have increased access to the benefits of biotechnological research. Plant scientists with the FAO require excellent graduating grades in a particular area of interest, such as new DNA techniques, molecular biology, or gene transfer technologies.

Careers in Focus

Sustainable Agriculture

The existing food system in the developed world is based on mass production. Consumers can choose what they want to eat at reasonable prices. At the same time, many consumers are concerned about the nutritional quality of mass-produced food and health risks from chemical inputs. There is also a feeling that food does not taste as good as it used to. Many farmers are unhappy with the intensive methods that they have to use in order to compete in the world market. Still, the global trend is for more industrialization of farming, not less. Industrialized farms produce more than 90 percent of the food in the developed world. While the percentage is much less in the developing world, it is increasing every year.

> **There is a feeling that food does not taste as good as it used to.**

Industrialized agriculture is based on economic factors. Large-scale farming is assumed to be more efficient than small-scale farming. Still, the economics of intensive farming tend to ignore related problems such as soil erosion, water contamination, pesticide poisoning, and waste disposal.

Sustainable farming conserves nutrients, soil, and water, while avoiding overuse of chemical inputs. It may not be able to match the yields of intensive farming, but it can be competitive in terms of overall efficiency. Those who support sustainable agriculture have to show that their methods can compete in the global marketplace. Can they produce food with similar variety and at a competitive price? Can they provide employment that does not return workers to the conditions of peasant economies?

THE IMPORTANCE OF SOIL

The word "soil" has come to be associated with dirt—something unclean and worthless. Yet soil supports life on Earth and is far more complex than it appears to be. Its fertility depends on millions of organisms. Nitrogen, phosphate, and potassium fertilizers are not enough to promote healthy plant growth. Many other nutrients, minerals, and microorganisms are required. A farmer who tills the soil by hand is likely to have a greater feeling for its life-giving properties than someone who works in an air-conditioned tractor. At the same time, returning to the use of hand-tools would clearly be impractical. Somehow, awareness of the importance of soil must be increased in order to preserve the world's plant-growing capability.

Soil is a nonrenewable resource that is endangered by deforestation, over-development, and chemical pollution.

The number of organic farms, which were the norm before 1940, is on the rise. Despite this renewed interest, only 10 percent of farms in the most environmentally friendly countries, such as Sweden, Denmark, Austria, and Germany, are organic. A more encouraging sign is that farmers are beginning to pay attention to sustainability. In the U.S., concern over soil erosion has changed the way many farmers plow their fields. Worldwide, wind, water, and poor farming practices contribute to soil erosion losses of 26 billion tons (24 billion t) each year. Erosion could be ended by the use of **no-till agriculture**.

There is a movement that favors planting a variety of crops in the same field. This is known as **intercropping**. This practice helps control pests. Numerous other ideas for sustainable farming are being developed,

The number of individual subsistence farms, or small-family mixed-farming operations, is decreasing in developed countries.

MAD COW DISEASE

Bovine Spongiform Encephalopathy (BSE), or mad cow disease, is a particularly unpleasant condition that rots the brains of cattle. First identified in Britain in the 1980s, it was affecting more than 1,000 cows a week by 1993. The common link seemed to be that the affected cows had eaten feed made partly from slaughterhouse waste. This contained cattle carcasses as well as the remains of sheep infected with scrapie, another brain-rotting disease. BSE caused a health scare in Britain when it was discovered that the disease may be able to jump from species to species. It is believed that BSE is linked to Creutzfeldt-Jakob disease in humans. Exports of British beef were banned and hundreds of thousands of cattle were slaughtered. Disposing of their carcasses has presented a problem. New European laws on landfill sites mean that they cannot be buried. One solution has been to burn the remains. High-temperature incineration kills any infectious proteins in the carcasses.

Sustainable Agriculture

mixing new technology with traditional know-how.

An important part of sustainability is mixed farming. When different crops are rotated or planted close to each other, weeds and insect pests are less of a problem than when single crops are planted. Instead of pesticides, the farmer can use natural predators to keep pests away. Today's farmers are much more aware of the dangers of pesticides and use them more carefully. Twelve pesticides that were once used in developed countries have now been banned or restricted in 78 countries.

In Indonesia, integrated pest management has become a national farm strategy. This strategy has reduced the amount that has to be spent on pesticides. Some farming communities in the developing world have been able to double or triple crop yields without using chemical inputs. Even in industrialized countries, farmers who switch to sustainable practices can enjoy higher profits. At first, average yields are likely to drop 10 to 20 percent. This makes it difficult for farmers to commit to sustainable practices. The transition must be made slowly. A single organic farmer surrounded by intensive farms may have his or her efforts ruined by chemical sprays drifting across the land. Water contamination is another potential problem as fertilizers and pesticides run into ditches, streams, and rivers that flow through neighboring farms. Also, it takes time to improve the nutrient levels of soil damaged by decades of heavy plowing, intensive crop production, and chemical application. Farmers need financial support to help them make the switch to more sustainable methods.

Today's farmers are much more aware of the dangers of pesticides.

The Amish community of Pennsylvania produces high-quality food using organic farming methods that do not rely on the use of modern technology. While most farmers would not go so far as to return to horse-and-wagon agriculture, the Amish are proof that sustainable farming can work.

The big companies that drive the food production industry are well aware of public concern about the damage done by intensive farming. They are keen to show that they are supporters of the sustainable approach.

KEY CONCEPTS

Community Supported Agriculture (CSA) CSA connects local farmers with local consumers. This relationship develops a strong regional food supply, which results in a stronger local economy. Supporters cover a farm's yearly operating expenses by purchasing a share of the season's harvest. Members help pay for seeds, fertilizer, water, labor, and equipment repair. In return, the farm provides fresh, high-quality, healthy produce. This agricultural model traces is roots back to Japan in 1970. A group of women concerned about the increase in food imports and decrease in the farming population initiated a direct growing and buying relationship between their group and local farmers. Called *teikei* in Japanese, it translates as "putting the farmers' face on food." The concept traveled first to Europe and then to North America, where about 1,000 CSA farms can be found across the United States and Canada.

Cassava, oranges, and tomatoes are the three most important cash crops for farmers in Ghana.

THE ORGANIC ALTERNATIVE

More than 90 percent of food in the developed world is produced by intensive methods using chemical fertilizers and pesticides. However, there are signs of change. There is growing evidence that over time organic and other sustainable methods can produce competitive yields at competitive prices. A new agricultural revolution may be underway, driven by consumer demand. Fruits, vegetables, grain, and livestock grown or raised on organic farms are produced "naturally" and with minimal impact on the environment. Rather than use pesticides and chemical fertilizers, organic farmers utilize nature to prevent disease and promote the growth of healthy plants and livestock. Compost and manure are among the natural substances used in place of chemical fertilizers. Insects, birds, and intercropping are among numerous ways in which disease is controlled and prevented. Livestock are fed organic foods, and are able to roam outside rather than being kept indoors in crowded conditions. Though given vaccinations and medically treated when ill, they are not given hormones and antibiotics to stimulate growth and prevent disease. Because organic foods are produced and processed without artificial ingredients or preservatives, they are generally thought to be of superior taste and quality. Organic produce requires more labor and is therefore more expensive, but if the environmental and health costs associated with current farming practices are taken into consideration, organic foods can be said to cost the same as, if not less than, food produced using more intensive methods.

Sustainable Agriculture

However, the multinationals have commercial interests in fertilizers and pesticides, too. It is difficult for them to favor a reduction in the use of these products—profits would fall. This is why sustainable practices are tough to put in place on a large scale. About 60 percent of agricultural research in the U.S. is privately funded. Very little of that research is directed at sustainable farming.

Community Supported Agriculture (CSA) is another trend visible in Europe, North America, and Japan. This approach links consumers directly to local farmers. CSA members help pay for the cost of running a farm and in return are provided with fresh quality produce throughout the growing season.

Food companies are market-driven. If current farming practices are to change, it will only be because consumers demand it.

THE FOOD AND AGRICULTURE ORGANIZATION (FAO)

Founded in 1945, the Food and Agriculture Organization is one of the largest agencies of the United Nations (UN). The FAO is the lead agency for agriculture, fisheries, forestry, and rural development. An intergovernmental organization, it has 183 member countries and functions on a budget of $650 million per year. The FAO's main aim is to eliminate hunger by improving agriculture in developing countries. It carries out hundreds of support projects all over the world. Since the early 1960s, FAO efforts have reduced the proportion of hungry people in the developing world from more than 50 percent to less than 20 percent. The organization encourages sustainable agriculture and rural development as long-term strategies for increasing food production and food security, while conserving natural resources. From assistance and information programs to international forums and advice to struggling governments, the FAO assesses the global structures of food and agriculture. It strives to ensure fair practices in the food trade and sets food standards and guidelines.

FAO Field Program Expenditures Per Major Activity in 1999
(percent of total expenditures)

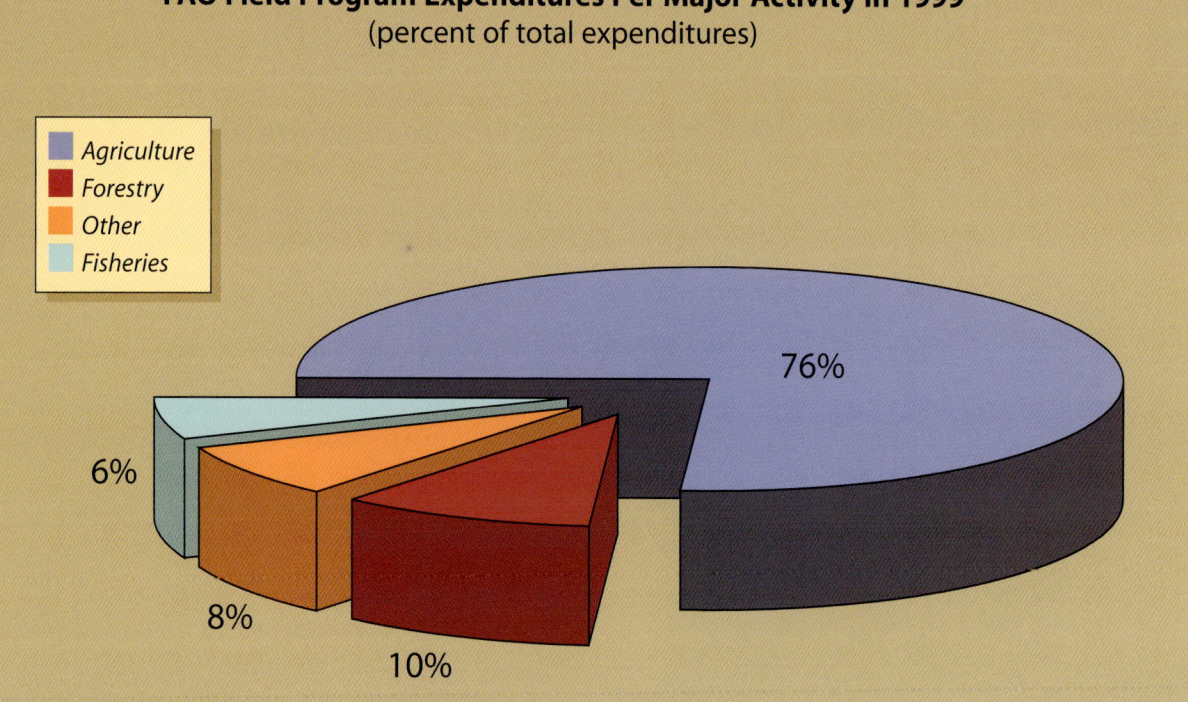

▬ Since fresh produce is sprayed with chemicals and pesticides, it is important to thoroughly wash all non-organic produce before eating it.

Sustainable Agriculture

Time Line of Events

8500 B.C.
People begin raising domesticated sheep and goats in the Middle East.

5000 B.C.
People in Central America begin growing corn.

2000 B.C.
Native Americans are cultivating corn, beans, squash, sunflowers, and other plants.

A.D. 400
The **serf** system of farm labor is established in Europe.

1612
Tobacco farming begins in Virginia.

1647
Rice cultivation begins in North and South Carolina.

1701
Jethro Tull invents the seed-drill, the first agricultural machine.

1727
Coffee plantations are established in Brazil.

1793
Eli Whitney patents the cotton-gin.

1850
The John Deere company manufactures 10,000 iron plows.

1870
Farmers account for less than 50 percent of workers in the U.S.

1900
Farm population in the U.S. is about 30 million.

1904
The gas-powered tractor is developed by Benjamin Holt.

1906
The Meat Inspection Act is passed.

1916
The U.S. Grain Standards Act is passed.

1917
The Ford Motor Company introduces the Fordson tractor.

1932
The first tree patent is awarded to James Markham.

1940
The self-tying hay baler is invented.

1943
The pesticide DDT is introduced in the U.S.

1950
Tractors outnumber horses on U.S. farms.

1953
The Agricultural Research Service is formed.

1957
The Humane Slaughter Act is passed.

1967
The Wholesome Meat Act is passed.

1969
The Food and Nutrition Service is established.

▬ Pivot irrigation, in which a sprinkler pivots around a central well, has enabled farming in arid and semi-arid regions.

1972
DDT is banned in the U.S.

1973
The California Certified Organic Gardeners association is formed.

1990
Farm population in the U.S. is 4.5 million.

1992
Agracetus receives a patent for its transgenic cotton.

1993
The Flavr Savr tomato is created.

1994
Farmers in the U.S. spend $16.4 billion on fertilizers and pesticides.

1996
The Federal Agricultural Improvement and Reform Act is passed.

1999
The USDA permits certain meat and poultry products to carry the label "certified organic."

2000
The USDA announces that certified organic cropland more than doubled in the 1990s.

Concept Web

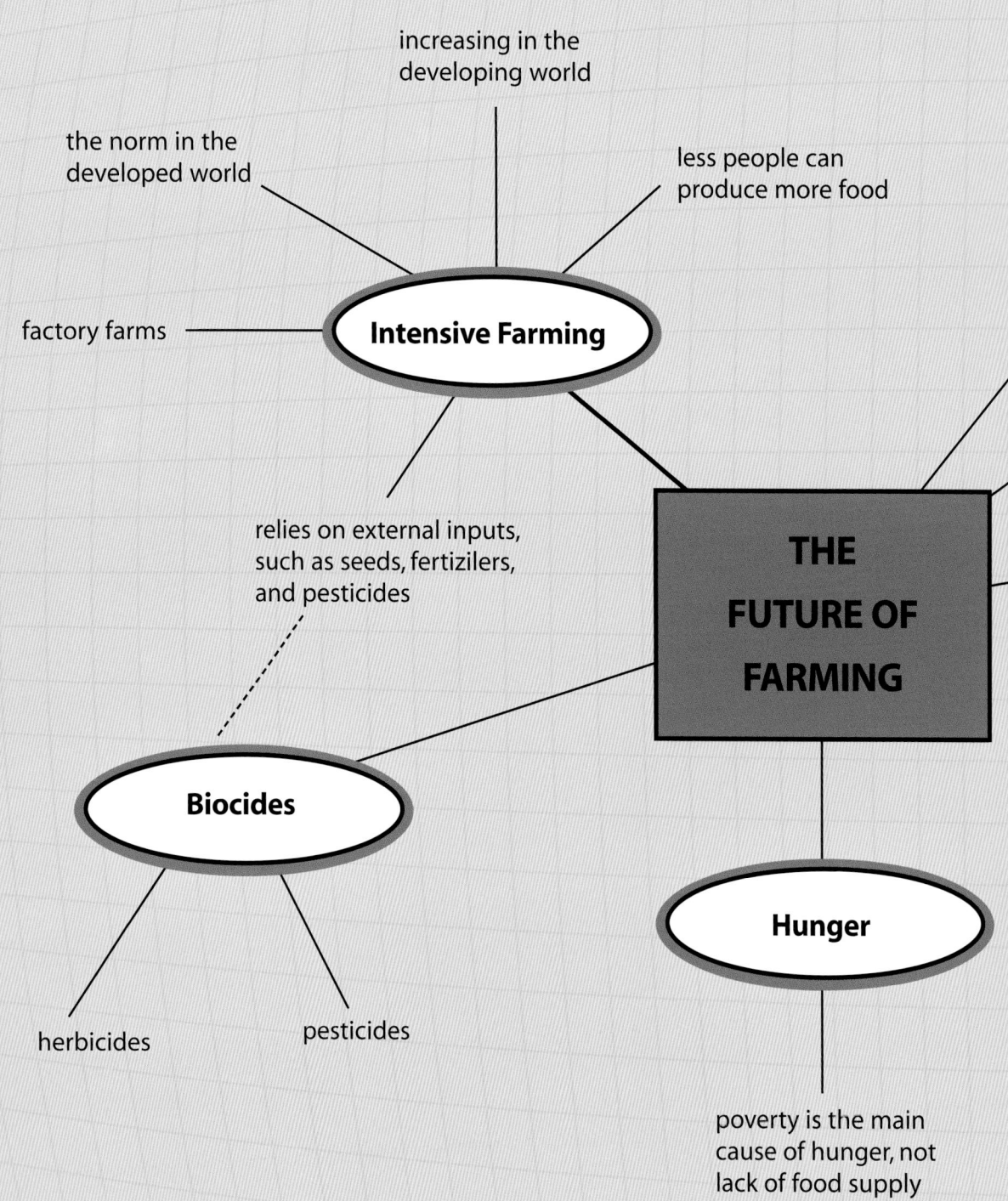

Glossary

biodegradable: capable of decomposing naturally

biotechnology: the use of living organisms to manufacture products

colonies: territories that are separate from but that are governed by a ruling country

crossbred: produced by the mating of different breeds or species

developed countries: those countries that have undergone the process of industrialization

developing world: those countries that are undergoing the process of industrialization, sometimes collectively referred to as the "Third World"

European Union: a political and economic union of 15 countries in Europe

excreta: waste matter discharged from the body

hemoglobin: a protein in red blood cells that transports oxygen from the lungs to the tissues

hormones: a substance produced in the body that controls growth and development

hybrid: in agriculture, a plant that is produced from a cross between two different plants in order to produce higher yields

insulin: a hormone that controls the level of glucose in the blood

intercropping: the planting of plants that deter pests among crops

irradiation: the use of radiation to preserve food

microorganisms: organisms such as bacteria and viruses so small that they can only be viewed under a microscope

nitrogen: a gas that makes up the majority of Earth's atmosphere; important for plant growth

no-till agriculture: a method of farming in which the soil is left undisturbed from harvest to planting

organic farming: farming without the use of synthetic fertilizers, pesticides, and drugs

patent: an exclusive right given to an inventor by the government to sell their product for a certain number of years

salmonella: a bacterium capable of causing food poisoning, often contracted through the consumption of undercooked meat

serf: an agricultural laborer in medieval Europe

soil erosion: the removal of the top layer of earth by natural forces, such as water, or wind

subsidies: financial aid given by the government to an individual, company, or other government

sustainable agriculture: method of farming that attempts to minimize damage to the environment

yields: amounts produced by cultivation or labor, especially of a crop

Further Reading

Charles, Daniel. *Lords of the Harvest: Biotech, Big Money, and the Future of Food.* Cambridge, MA: Perseus Books, 2001.

Ekarius, Carol. *Small-Scale Livestock Farming: A Grass-Based Approach for Health, Sustainability, and Profit.* North Adams, MA: Storey Books, 1999.

Perles, Catherine. *The Early Neolithic in Greece: The First Farming Communities in Europe.* Cambridge: Cambridge University Press, 2001.

Pinstrup-Andersen, Per, and Ebbe Schioler. *Seeds of Contention: World Hunger and the Global Controversy Over GM (Genetically Modified) Crops.* Baltimore: Johns Hopkins University Press, 2001.

Schwenke, Karl. *Successful Small-Scale Farming: An Organic Approach.* North Adams, MA: Storey Books, 1991.

Answers

Multiple Choice
1. d) 2. d) 3. b) 4. d) 5. a) 6. a) 7. c)

Where Did It Happen?
1. The Middle East
2. Kansas, United States
3. Brazil and Malaysia
4. Pennsylvania, United States

True or False
1. F
2. T
3. T
4. T

Internet Resources

The following Web sites provide more information on farming:

USDA
http://www.usda.gov
The USDA's Web site provides information and resources on the issues, goals, and structure of the United States Department of Agriculture. Visitors can browse the agencies, services, and programs to discover the breadth of the USDA. The site is updated on a regular basis, providing current news and stories related to the world of agriculture. For those interested in reading in-depth studies and reports on agriculture-related topics, such as biotechnology, government authorized links are provided.

FAO
http://www.fao.org
The FAO's Web site has lots of important information related to the world's food sources. With a focus on sustainable development, the FAO's Web site offers visitors such categories as agriculture, economics, fisheries, forestry, and nutrition. The "News and Highlights" section offers current articles along with access to archived information. "Global Watch" lists stories and reports related to impending food availability problems. For instance, visitors can educate themselves on the effect wars and famines are having on food production and supply across the globe.

Some Web sites stay current longer than others. To find other farming Web sites, enter terms such as "mixed farming," "organic farming," or "pesticides" into a search engine.

6. Approximately what percent of the world's land area is used to grow crops?
   a) 11 percent
   b) 16 percent
   c) 23 percent
   d) 28 percent

7. What is bush meat?
   a) Meat that comes from animals raised on organic farms
   b) Meat that comes from animals raised on factory farms
   c) Meat that comes from wild animals in African forests
   d) None of the above

Where Did It Happen?

1. Sheep and goats were first raised here in 8500 B.C.
2. The location of a movement to plant a variety of crops in the same field
3. Rubber occurs naturally in these two countries
4. The Amish have been producing organic foods here without the use of mechanization for more than a century

True or False

1. The number of people making a living from agriculture has increased since the Industrial Revolution.
2. The biggest change brought about by intensive farming is that a small number of companies provide the basics, such as seeds and fertilizers, to the entire agriculture industry.
3. Mixed farms produce both livestock and crops.
4. About 25 percent of the land on Earth is permanent pasture.

Answers on page 53

Quiz

Multiple Choice

1. The world's population is expected to reach 9.5 billion by 2050. In order to meet the demand for food by that year, experts say the global farm output will have to increase by:
    a) 10 percent
    b) 20 percent
    c) 30 percent
    d) 300 percent

2. Overuse of pesticides to protect crops can:
    a) eliminate insects and other pests for good
    b) pose a threat to humans, animals, and the environment
    c) lead to an increase in the pests' resistance
    d) both b and c

3. A transgenic animal is an animal that:
    a) has been imported from another country
    b) contains the genes of a foreign species
    c) does not have the ability to reproduce
    d) none of the above

4. What percent of the European Union's farm support money goes toward environmentally friendly farming?
    a) 10 percent
    b) 30 percent
    c) 7 percent
    d) 2 percent

5. Many multinational farming companies have operations in countries where food production and labor is cheaper than in the developed world. This approach:
    a) pressures farmers in the developed world to turn to intensive farming to compete
    b) eases the strain on farmers in the developed world to produce large yields
    c) allows multinationals to learn about and employ local, traditional farming methods
    d) frees up land in the developed world for government parks and wilderness areas

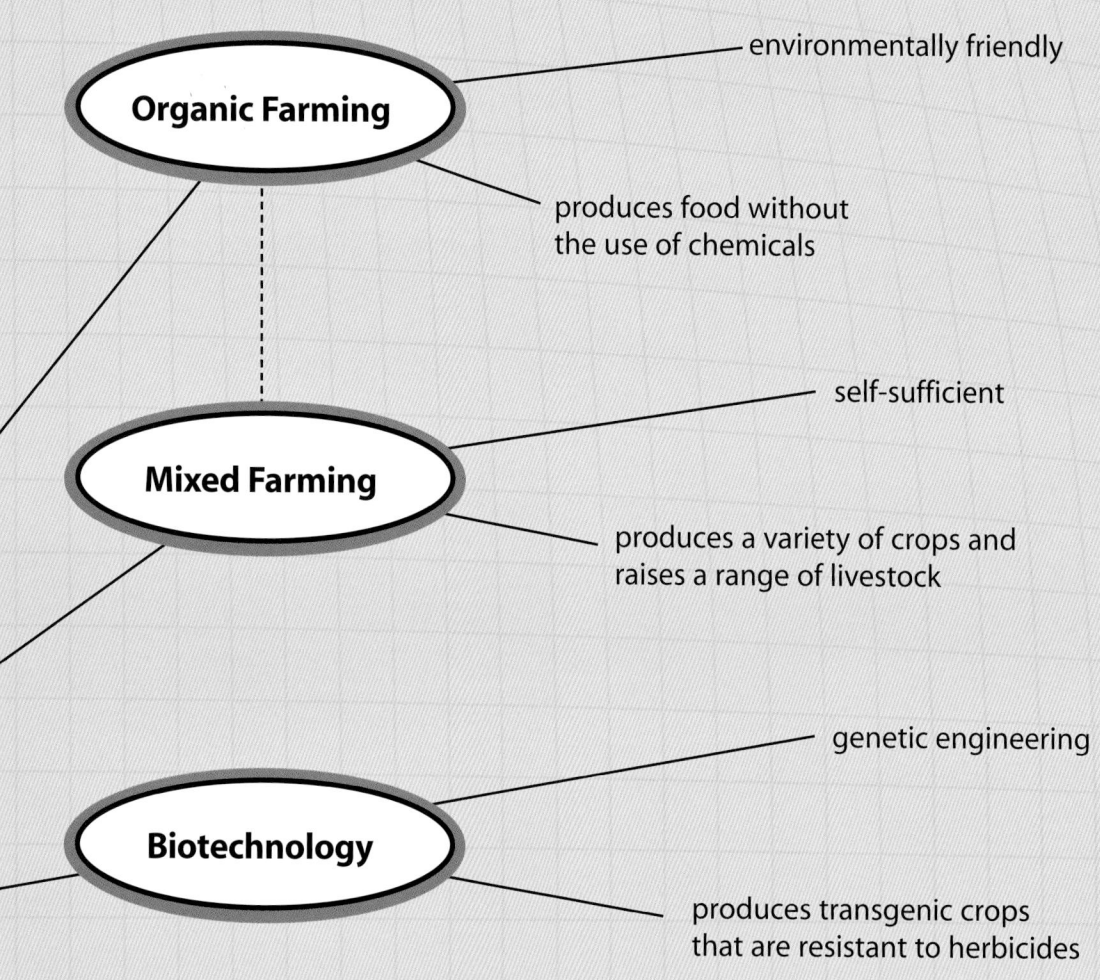

MAKE YOUR OWN CONCEPT WEB

A concept web is a useful summary tool. It can also be used to plan your research or help you write an essay or report. To make your own concept web, follow the steps below:
- You will need a large piece of unlined paper and a pencil.
- First, read through your source material, such as *The Future of Farming* in the Understanding Global Issues series.
- Write the main idea, or concept, in large letters in the center of the page.
- On a sheet of lined paper, jot down all words, phrases, or lists that you know are connected with the concept. Try to do this from memory.
- Look at your list. Can you group your words and phrases in certain topics or themes? Connect the different topics with lines to the center, or to other "branches."
- Critique your concept web. Ask questions about the material on your concept web: Does it all make sense? Are all the links shown? Could there be other ways of looking at it? Is anything missing?
- What more do you need to find out? Develop questions for those areas you are still unsure about or where information is missing. Use these questions as a basis for further research.

Index

Amish 42
animal rights 20, 24
antibiotics 23, 24, 43
Asia 7, 8, 11, 18, 28, 29

biocides 17, 38, 48
biodegradable 17, 35
biodiversity 36
biotechnology 5, 35, 36, 37, 39, 49
broiler chickens 21, 24
bush meat 22

colonies 11
Community Supported Agriculture (CSA) 42, 44
consumers (grocery shoppers) 5, 7, 20, 31, 40, 43, 44
crops 5, 9, 11, 12, 13, 14, 15, 16, 17, 18, 19, 26, 31, 35, 36, 37, 38, 41, 42, 43, 49

DDT 17, 46, 47
developed world 5, 8, 11, 12, 14, 17, 18, 20, 24, 30, 31, 33, 35, 40, 41, 42, 43, 48
developing world 7, 8, 14, 18, 30, 33, 35, 36, 39, 40, 42, 44
diseases 17, 23, 24, 36
DNA 35, 36, 37, 39

environment 5, 8, 9, 13, 14, 15, 19, 24, 31, 32, 37, 38, 41, 43, 49
European Union 7, 31

factory farms 5, 9, 48
farmers 5, 7, 8, 11, 12, 13, 14, 15, 17, 18, 19, 20, 23, 26, 31, 32, 33, 35, 36, 37, 39, 40, 41, 42, 43, 44, 46, 47
fast-food industry 8
fertilizers 5, 14, 15, 16, 17, 18, 19, 30, 40, 42, 43, 44, 48

Flavr Savr 31, 47
Food and Agriculture Organization (FAO) 8, 39, 44
food safety 5, 24
fruits 9, 16, 26, 30, 31, 35, 43

genetic engineering (genetic modification) 8, 12, 14, 31, 35, 37, 38, 39, 49
globalization 30, 33
grain trade 18, 30
growth hormones 21

herbicides 36, 38, 48, 49
hybrid 17, 18

industrialized agriculture 5, 7, 8, 11, 14, 21, 23, 32, 40
Industrial Revolution 11, 12, 18, 30
intensive farming 8, 11, 12, 14, 16, 19, 22, 23, 33, 38, 40, 42, 43, 48
intercropping 41, 43

livestock 5, 7, 11, 13, 18, 20–25, 36, 38, 43, 49

mad cow disease (BSE) 41
meat 5, 6, 7, 8, 11, 22, 23, 24, 25, 30, 36
microorganisms 12, 18, 35, 40
mixed farming 11, 13, 42, 49
multinationals 30, 32, 33, 42

organic farming 5, 19, 42, 43, 49

patents 35, 38, 46, 47
pesticides 5, 9, 12, 14, 17, 18, 19, 29, 30, 37, 42, 43, 44, 45, 46, 47, 48
pests 9, 12, 13, 16, 17, 18, 35, 37, 38, 39, 41, 42
plantations 11, 12, 13, 46
population growth 8, 28

ranching 25
refrigeration 30

soil 12, 13, 14, 15, 17, 18, 19, 24, 28, 32, 36, 40, 41, 42
subsidies 31, 32
sustainable agriculture 5, 19, 25, 40, 41, 42, 43, 44

transgenic 35, 36, 37, 38, 47, 49

United States 7, 11, 12, 14, 18, 21, 23, 24, 29, 30, 32, 35, 36, 41, 42, 44, 46, 47

vegetables 5, 7, 9, 16, 26, 30, 31, 35, 43
vegetarian 5, 20, 24

Photo Credits

Cover: Tomatoes on Vine (**Corbis**); **Title Page**: Digital Stock; **Kindra Clineff Photography**: pages 2/3; **Corbis**: page 34; **CORBIS/MAGMA**: page 21 (©**David Reed**); **DigitalVision**: pages 14, 22; **Digital Stock**: pages 6, 9, 10, 18, 19, 20, 30, 31, 38, 41, 45, 47; **Victor Englebert**: pages 13, 16, 43T; **EyeWire Inc.**: page 4; **Mike Grandmaison**: pages 23, 40, 43B; **Photo Agora**: page 33 (**Sean Sprague**); **PhotoDisc**: pages 15, 25, 39; **Still Pictures**: page 36 (**Nigel Dickinson**); ©**Ray Witlin/World Bank Photo**: page 7.